curious about

RECYCLING

BY AMY S. HANSEN

AMICUS LEARNING

What are you

curious about?

Curious About is published
by Amicus Learning,
an imprint of Amicus.
P.O. Box 227
Mankato, MN 56002
www.amicuspublishing.us

Editor: Alissa Thielges
Series Designer: Kathleen Petelinsek
Book Designer and Photo Researcher: Emily Dietz

Library of Congress Cataloging-in-Publication Data
Names: Hansen, Amy, author.
Title: Curious about recycling / by Amy S. Hansen.
Description: Mankato, MN: Amicus Learning, [2025] | Series: Curious about green living | Includes bibliographical references and index. | Audience: Ages 5–9 | Audience: Grades 2–3 | Summary: "Why is recycling important? Ignite kids' growing curiosity about the environment with an inquiry-based approach to reducing waste. Includes infographics and back matter to support research skills, plus table of contents, glossary, and index"—Provided by publisher.
Identifiers: LCCN 2023043284 (print) | LCCN 2023043285 (ebook) | ISBN 9781645496977 (library binding) | ISBN 9781681529660 (paperback) | ISBN 9781645497035 (ebook)
Subjects: LCSH: Recycling (Waste, etc.)—Juvenile literature. | Waste minimization—Juvenile literature.
Classification: LCC TD794.6 .H36 2025 (print) | LCC TD794.6 (ebook) | DDC 363.72/82—dc23/eng/20231214
LC record available at https://lccn.loc.gov/2023043284
LC ebook record available at https://lccn.loc.gov/2023043285

Photo Credits: Alamy: Jim West 10; Dreamstime: Aleksey Popov 2, 13 (r), Aliaksei Shaviakou 14–15, Grigorenko 5, Oksana Krasiuk 3, 17; Getty: Aslan Alphan 16, Edwin Tan 18, LordHenriVoton 21, -VICTOR- 19 (bulb); iStock: Jun cover, 1; Noun Project: Graphixs_Art PK 19, 22 & 23 (can), lipi FR 13 (recycling symbol), Weno NL 22 & 23 (recycling icons); Shutterstock: BAZA Production 8, inspiring.team 9, 12, 13 (icons), ivector 9 (icons), Katerina Primula 13 (egg carton), Part of Design 11, SeventyFour 2, 9 (l), Skylines 4, Sudowoodo 9 (t), Vechterova Valeria 6–7

CHAPTER ONE

What should I do with an empty can?

Place cans in the right bin so they can be recycled.

Stop! Don't put that can in the trash. Put it in the **recycling** bin. It has a long life ahead of it. Recycling is reusing **raw materials**. The **aluminum** in that can will make a new can. Or it will make something brand new!

Cans are crushed into blocks at recycling centers.

DID YOU KNOW?
It takes about 60 days to recycle a soda can.

Why should I recycle?

Recycling 1 ton (0.9 metric ton) of paper saves 17 trees.

It helps save **resources**. Recycling paper saves trees. Recycling cans saves metals. It also saves space at the **landfill**. And it saves water and energy. It takes less of those to make recycled **products** than new products.

Recycling bins are a different color than garbage bins.

What can I recycle?

Lots of things! You can recycle old worksheets. The cardboard box from your new shoes can be recycled, too. Add your juice box and milk jug. Ask an adult if there are things you aren't sure about. Check the services in your area.

Make sure to rinse the containers before recycling them.

A CLOSER LOOK

CHAPTER TWO

Where does it go after the recycling bin?

A forklift stacks blocks of paper while plastic jugs travel up a conveyer belt.

Your bin gets dumped into a truck. The truck goes to a recycling center. The center separates the items. Paper goes one way. Glass goes another way. The **plastics** get shredded. The metals are crushed and melted. They form giant blocks. Companies buy the blocks and make new products.

What do the numbers on plastic mean?

These tell you what kind of plastic it is. The numbers are on the bottom of a container. A milk jug has a number two. People at the recycling center sort the plastic by numbers. Then the plastic is melted to make something new.

TYPES OF PLASTIC

A machine melts one type of plastic all together.

Hard to Recycle

Easy to Recycle

Hard to Recycle

Very Hard to Recycle

A CLOSER LOOK

What's so hard about recycling plastic?

DID YOU KNOW?
Most plastic is made from fossil fuels. These harm the air. Some new plastics are now made from vegetable oil.

There are many types of plastic. And none of them can be melted together. It costs a lot to separate all of them. The material gets weaker, too. It can only be recycled one or two times. Then it gets thrown away.

Plastics are separated by color before melting.

TAKE ACTION

CHAPTER THREE

How do I know what to recycle?

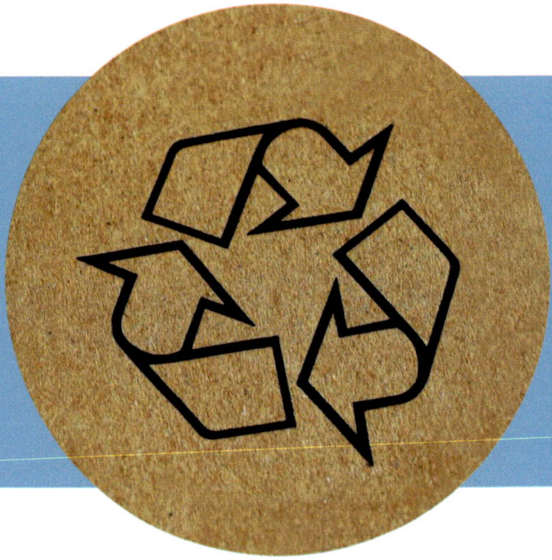

The recycling symbol is used around the world.

Look for the triangle symbol! That tells you if the item can be recycled. Watch out for items made of multiple materials. A toy's packaging could have both plastic and cardboard. Each can be recycled. But they should be broken apart to make it easier.

DID YOU KNOW?
Gary Anderson made the recycling symbol in college. He won a contest in 1970. Now his symbol is everywhere.

A boy helps his mom separate the recycling materials.

What if I can't recycle?

Drinking from a reusable water bottle can save on plastic.

Not every town or city has a recycling center. But you can still help. Be careful of what you throw away. Use both sides of a piece of paper. Carry a refillable water bottle if you can. Save cans and take them when you travel.

DID YOU KNOW?

Recycling saves energy. It takes 40 percent less energy to make a bottle from recycled glass than from new glass.

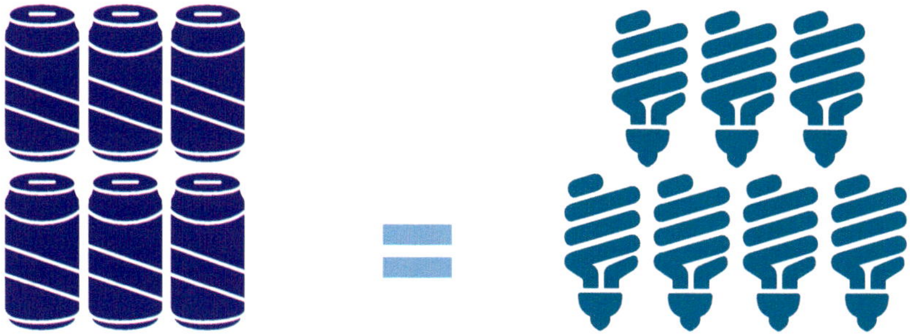

60,000 recycled cans = **14,000 kilowatt-hours of energy saved**

This is more than enough energy to power the lights in your school for a year.

How can we get more people to recycle?

Some states pay people five cents a can. This makes more people want to recycle. People look through garbage. They find cans and bottles. A boy from Hawaii has collected more than 1 million cans and bottles. He's raised about $50,000! He uses the money to help kids go to college. Can you think of other ideas?

A young woman brings bottles to a city collection bin.

ASK MORE QUESTIONS

What happens to paper when it is recycled?

Where is the closest recycling center to my home?

Try a BIG QUESTION: What would make recycling important to more people?

SEARCH FOR ANSWERS

Search the library catalog or the Internet.
A librarian, teacher, or parent can help you.

Using Keywords
Find the looking glass.

Keywords are the most important words in your question.

If you want to know:

- what happens to recycled paper, type: RECYCLING PAPER PROCESS
- where the closest recycling center is, type: RECYCLING CENTER NEAR ME

LEARN MORE

FIND GOOD SOURCES

Are the sources good?
Some are better than others. An adult can help you. Here are some good, safe sources.

Books

Life of a Soda Can by Louise Nelson, 2023.

Plastic by Trudy Becker, 2024.

Reduce, Reuse, Recycle by Nick Rebman, 2022.

Internet Sites

Ready, Set, Sort!
http://rhodeislandresource.recycle.game/
Learn how to sort different items in this interactive game.

Tour a Recycling Center
https://kids.nationalgeographic.com/science/article/tour-a-recycling-center
National Geographic explores the planet and its problems. It is a good source for animals and nature.

Every effort has been made to ensure that these websites are appropriate for children. However, because of the nature of the Internet, it is impossible to guarantee that these sites will remain active indefinitely or that their contents will not be altered.

SHARE AND TAKE ACTION

Collect litter from a park.
Separate the trash from the recycling.

Hold a contest with your friends.
Who can collect the most cans or bottles to recycle?

Create a poster.
List what items can and can't be recycled in your area. Put the poster up at your school to remind people to recycle.

GLOSSARY

aluminum A silver metal that is strong and light.

fossil fuel A fuel that is formed in the ground from dead plants or animals.

landfill An area where waste is buried under the ground.

product Something that is made or grown to be sold or used.

plastic A light, strong substance made with oil that can be made into different shapes.

raw material A basic material that can be used to make or create something.

recycle To process old items made of glass, plastic, paper, and metal so they can be used to make new products.

resource Something valuable or useful to a place or person.

INDEX

About the Author

Amy S. Hansen lives in Maryland. She started writing as a reporter for local newspapers. After a few years she went back to school to study environmental science and then worked as a science reporter. Nowadays, she (mostly) writes for kids because it is much more fun.